INTERNATIONAL GCSE (9–1)

NICK ENGLAND
NICKY THOMAS

Physics

for Edexcel International GCSE

WORKBOOK

HODDER
EDUCATION
Learn more

Orders: please contact Hachette UK Distribution, Hely Hutchinson Centre, Milton Road, Didcot, Oxfordshire, OX11 7HH. Telephone: +44 (0)1235 827827. Email education@hachette.co.uk Lines are open from 9 a.m. to 5 p.m., Monday to Friday. You can also order through our website: www.hoddereducation.co.uk

© Nicky Thomas 2017

First published in 2017 by
Hodder Education,
An Hachette UK Company
Carmelite House
50 Victoria Embankment
London EC4Y 0DZ

www.hoddereducation.co.uk

Impression number 10

Year 2024

Illustrations by Aptara

Typeset in Frutiger 55 roman 10/13 pt by Integra Software Services Pvt. Ltd., Pondicherry, India

Printed in the UK

A catalogue record for this title is available from the British Library.

ISBN: 978 1510 40566 0

Contents

Preface

Physics for Edexcel international GCSE Workbook is the new edition of the Edexcel International GCSE Physics Practice Book. It is designed as a 'write-in' workbook for students to practise and test their knowledge and understanding of the content of the International GCSE Physics course.

The sections are presented with the same headings and in the same order as in the Student Book, *Edexcel International GCSE Physics Student Book Second Edition*.

The Workbook should be used as an additional resource throughout the course alongside the Student Book. The 'write-in' design is ideal for use by students in class or for homework.

Answers can be found online at www.hoddereducation.co.uk/igcsephysics

1 Forces and motion

1 A group of students close a door using a newton meter to measure the force. Their results vary from 2.5 N to 3.2 N. Suggest **two** reasons why their results are not all the same. *[2 marks]*

....

....

....

(Total = 2 marks)

2 A student investigates whether the time for a ball bearing to fall 50 cm through oil depends on the ball's mass.

a) State **two** variables the student should measure. *[2 marks]*

....

....

b) Explain **one** reason why the tube of oil is more suitable than a measuring cylinder for the experiment. *[2 marks]*

....

....

....

c) Suggest **one** reason why the experiment may not give valid results. *[2 marks]*

....

....

....

(Total = 6 marks)

3 A ball is dropped from a height of 3.0 m. How long does it take to reach the ground? Circle the correct answer. *[1 mark]*

A 0.30 s

B 0.34 s

C 0.77 s

D 0.60 s

(Total = 1 mark)

4 The speed–time graph shows how the speed of a plane taking off increases as it accelerates along the runway.

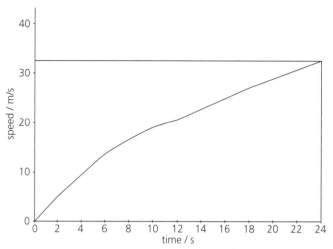

a) State Newton's second law of motion. *[1 mark]*

....

b) Use the graph to calculate the acceleration of the plane in the first 4 s. *[3 marks]*

....

....

....

....

c) Use the graph to calculate the average acceleration of the plane between 10 s and 20 s. *[3 marks]*

....

....

....

....

d) The forward force from the engine remains constant throughout the take-off.

Explain why the acceleration of the plane gets less as it speeds up. *[2 marks]*

....

....

....

e) The plane takes off when its speed reaches 30 m/s. How far has it travelled when it takes off? Use the graph above to calculate the distance, in m, then circle the answer closest to the correct distance. *[1 mark]*

A 30 m   B 100 m   C 350 m   D 600 m

f) The diagram below shows the plane flying at a constant speed and constant height.

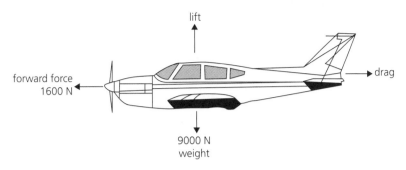

Which of the following rows in the table is correct? Circle the correct answer, choosing from A to D.   [1 mark]

	Lift force / N	Drag force / N
A	9000	1500
B	9050	1600
C	9050	1500
D	9000	1600

g) Calculate the total mass of the plane and its passengers. The gravitational field strength is 10 N/kg.   [2 marks]

....

....

(Total = 13 marks)

5   A ball is dropped 1.5 m and reaches the ground, travelling at 5 m/s.

a) If the ball's mass is 0.5 kg, calculate its momentum when it lands on the ground.   [3 marks]

....

....

b) When the ball lands, it takes 0.2 s to stop. Calculate the force, in N, which acts on the ball when it decelerates.   [3 marks]

....

....

....

c) Use your ideas about momentum to explain whether the ball experiences a greater force when it lands on wood or when it lands on carpet.   [3 marks]

....

....

(Total = 9 marks)

6 The diagram shows how the stopping distance of a car depends on the tread depth of the tyre. The tread is the pattern of grooves on the surface of a tyre.

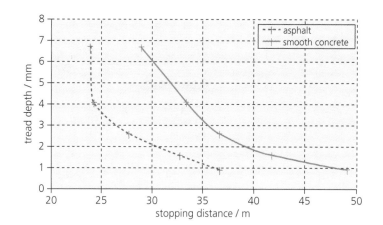

a) A student concludes the depth of the tyre tread affects the stopping distance more than the road surface. Evaluate her conclusion. *[3 marks]*

....

....

....

b) Suggest a suitable tyre tread depth using your ideas about stopping distances and friction. *[2 marks]*

....

....

(Total = 5 marks)

7 The graph shows how the speed of a vehicle affects the chance of a pedestrian being hurt. A pedestrian is more likely to receive a head injury when the front of a car hits them.

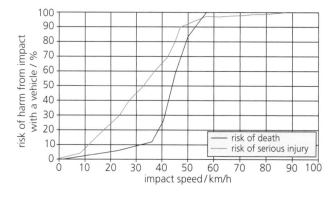

Paolo says: 'Installing speed cameras will reduce injuries most.'

Sarah says: 'Redesigning the front of cars will reduce injuries most.'

Use your ideas about forces and momentum to explain which student you agree with. *[4 marks]*

....

....

....

....

....

....

(Total = 4 marks)

8   The graph shows the stopping distance for vehicles travelling at the same speed on different road surfaces.

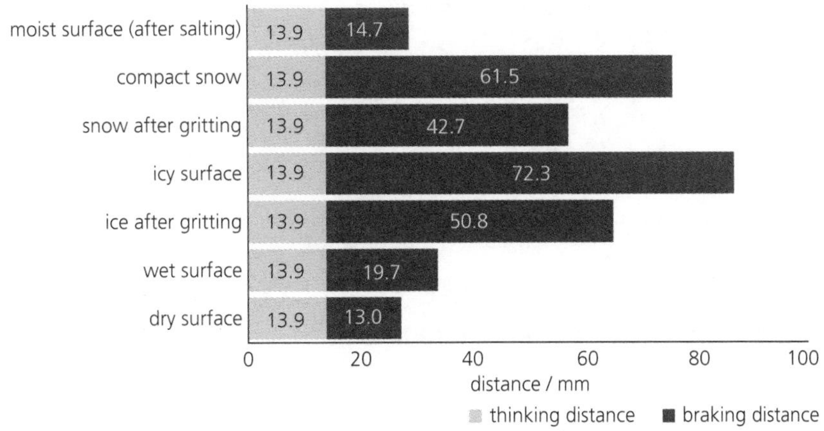

Using the graph and your knowledge of physics, state whether the speed limit should be higher or lower on roads that have not been gritted or salted. Explain your answer.   *[4 marks]*

....

....

....

....

....

....

(Total = 4 marks)

9   A student uses a metre ruler on a triangular pivot to find the weight of a small piece of wood. The student balances a 1 N weight on one side and the piece of wood on the other side of the pivot.

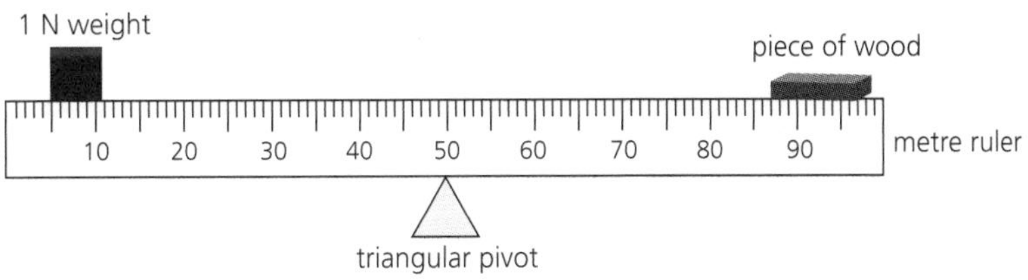

a) State the equation linking moment, force and distance from pivot.   *[1 mark]*

....

b) Explain how the student can use this equipment to find the unknown mass of the wood.   *[4 marks]*

....

....

....

....

....

c) Describe **two** sources of error in this experiment.   *[2 marks]*

....

....

....

(Total = 7 marks)

10 The diagram shows a spanner being used to turn a nut.

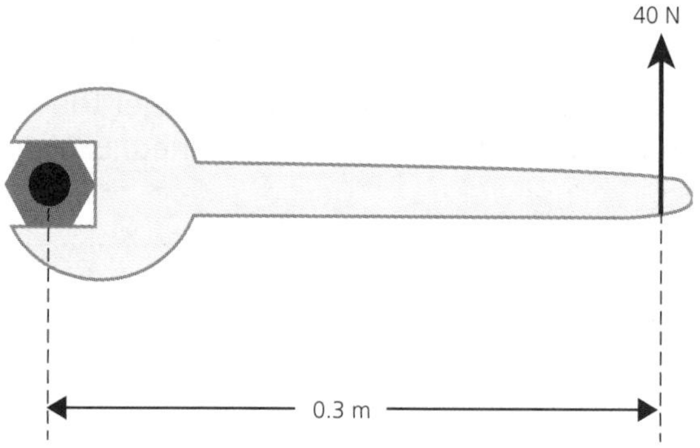

a) Calculate the turning moment of the force. State the unit.   *[2 marks]*

....

....

b) State **two** pieces of information needed to calculate whether the spanner can turn the nut. Give a reason for your answer.   *[3 marks]*

....

....

....

(Total = 5 marks)

11 This apparatus is used to measure the extension of a spring when a force is applied.

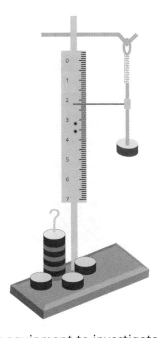

Explain how a student should use this equipment to investigate how the extension of the spring varies with applied force. [5 marks]

....

....

....

....

....

....

....

(Total = 5 marks)

12 a) Explain the difference between a vector quantity and a scalar quantity. [2 marks]

....

....

b) State whether the following quantities are vectors or scalar:

i) mass .... [1 mark]

ii) time .... [1 mark]

iii) velocity .... [1 mark]

iv) force. .... [1 mark]

(Total = 6 marks)

② Electricity

1 This multimeter is used to measure resistance. What is the reading? Include the unit.   *[2 marks]*

....

....

(Total = 2 marks)

2 A student uses a voltmeter in an experiment. What is the reading on the voltmeter?   *[2 marks]*

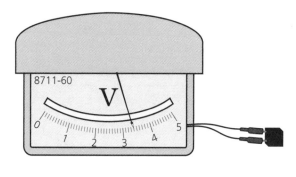

....

....

(Total = 2 marks)

3 A teacher sets up a van de Graaff generator.

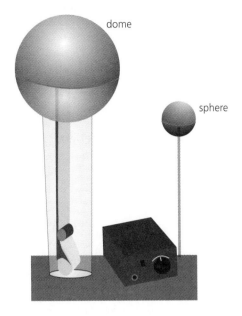

When the Van de Graaff generator is switched on, it creates a high voltage in the large metal dome. The metal sphere is close to the large metal dome so sparks move from the dome to the sphere.

Explain **two** safety precautions a teacher should take before using the Van de Graaff generator. *[4 marks]*

....

....

....

....

....

(Total = 4 marks)

4 A student was given this equipment: two lamps, connecting leads, a battery pack, three switches.

Draw the circuit that allows both lamps to be controlled at the same time and to be controlled independently. *[4 marks]*

(Total = 4 marks)

5 A student uses a multimeter and a light-dependent resistor (LDR) to investigate how the light level from a lamp varies as the distance from the lamp increases.

a) State **two** variables the student should control. *[2 marks]*

....

....

b) Describe **one** way the student can change the independent variable, including details of the equipment used. *[3 marks]*

....

....

....

....

(Total = 5 marks)

6 A teacher shows the class an oscilloscope, which is connected to an a.c. supply. The trace is shown below. The scale on the y-axis is set in volts/division.

a) What is the setting of the volts/division button on the oscilloscope? *[1 mark]*

....

b) What is the peak reading on the oscilloscope? *[1 mark]*

....

....

c) The teacher disconnects the a.c. power supply and connects a 6 V d.c. power source (a battery pack) instead. Describe the new trace shown on the screen. *[2 marks]*

....

....

....

(Total = 5 marks)

7 The diagram shows a 12 V battery in series with an ammeter, a variable resistor and a lamp.

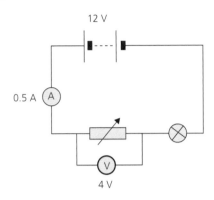

12 V

0.5 A Ⓐ

Ⓥ
4 V

The resistance of the variable resistor is **reduced**. Which of the following rows in the table is correct? Circle the correct answer, choosing from A to D. *[1 mark]*

	Reading on ammeter	Reading on voltmeter
A	larger	larger
B	larger	smaller
C	smaller	larger
D	smaller	smaller

(Total = 1 mark)

8 A 60 W bulb is connected to the mains supply, 230 V. Calculate the charge that flows through the bulb in 5 minutes, in coulombs, and circle the answer that is closest to the correct charge below. *[1 mark]*

A 300 C

B 1.3 C

C 18 000 C

D 78 C

(Total = 1 mark)

9 A student sets up the circuit shown in the diagram.

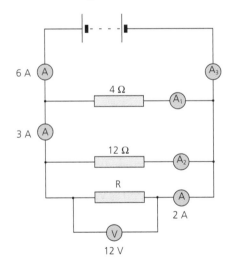

6 A Ⓐ   Ⓐ₃

4 Ω   Ⓐ₁

3 A Ⓐ

12 Ω   Ⓐ₂

R   Ⓐ
2 A

Ⓥ
12 V

a) Calculate the values of current recorded by the ammeters. *[3 marks]*

A₁ reading = ... A

A₂ reading = ... A

A₃ reading = ... A

b) i) State the equation that links voltage, current and resistance. *[1 mark]*

....

   ii) Calculate the value of the resistance, R. State the unit of resistance. *[2 marks]*

....

....

....

c) State the voltage of the battery. Explain your answer. *[2 marks]*

....

....

....

(Total = 8 marks)

10 A lamp and a heater are designed to work from a 230 V mains supply.

The lamp is labelled 230 V 11 W.

The heater is labelled 230 V 8 A.

a) i) State the current flowing through the heater during normal use. *[1 mark]*

....

   ii) State the energy transferred per second when the lamp is turned on. *[1 mark]*

....

   iii) The following fuses are available: 3 A, 5 A and 13 A.

      Explain why you should choose a 13 A fuse for the heater. *[1 mark]*

....

b) i) State the equation that links voltage, current and power. *[1 mark]*

....

   ii) Calculate the current in the lamp when it is working from a 230 V mains supply. *[1 mark]*

....

   iii) Explain which fuse you should choose for the lamp. *[1 mark]*

....

c) i) Show that the resistance of the heater is 28.75 Ω. *[2 marks]*

....

....

ii) Calculate the energy transferred by the heater if it works for 3 hours. State the unit of energy. *[3 marks]*

....

....

....

d) You take the heater to a country where the mains voltage is 110 V.

i) Calculate the current in the heater when it is connected to the 110 V mains supply. *[2 marks]*

....

....

ii) Calculate its power when it is connected to the 110 V mains supply. *[2 marks]*

....

....

(Total = 15 marks)

11 A student uses this circuit to investigate how the current in a diode changes with the voltage across it.

The graph shows the results of this investigation.

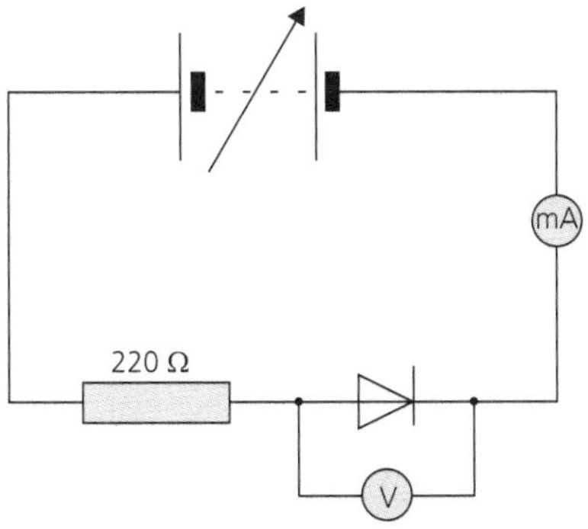

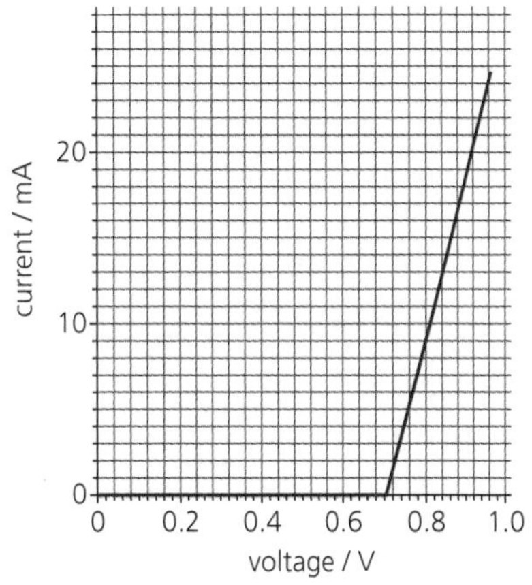

a) Describe how the current changes with the applied voltage. *[2 marks]*

....

....

b) Use the graph to calculate the voltage across the diode when the current is 10 mA. *[1 mark]*

....

c) Calculate the voltage across the resistor when the current is 10 mA.   [3 marks]

....

....

....

....

d) Use your answers to b) and c) to calculate the battery voltage. Explain your answer.   [2 marks]

....

....

....

(Total = 8 marks)

12 The circuit diagram is used to investigate the way the current in a filament lamp depends on the applied voltage. The results of the investigation are shown in the table.

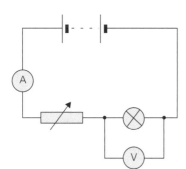

Current / A	0	0.5	0.7	1.2	1.7	2.1	2.3	2.8	3.1
Voltage / V	0	0.3	0.8	1.8	3.7	5.0	6.5	9.0	11.0

a) Explain why the current is the dependent variable in this investigation.   [1 mark]

....

....

....

b) Plot a graph of current (*y*-axis) against voltage (*x*-axis). Draw a line of best fit through the points.   [5 marks]

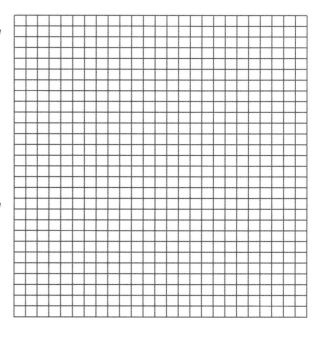

c) The student made an error with one ammeter reading.

Identify this error, write it below, and explain how to use the graph to calculate the correct current value. *[2 marks]*

....

....

d) The student extends the line of best fit to find the current value for 12.0 V. Explain why this value is unreliable. *[2 marks]*

....

....

e) Explain the shape of the line of best fit. *[2 marks]*

....

....

(Total = 12 marks)

13 The circuit and apparatus shown in the diagram are used to investigate how the resistance of a resistor, R, changes with temperature. The resistor's temperature is controlled by placing it in a beaker of water.

The results for a thermistor and a coil of wire are plotted on to a graph and a line of best fit is drawn for each set of results. The lines are labelled A and B.

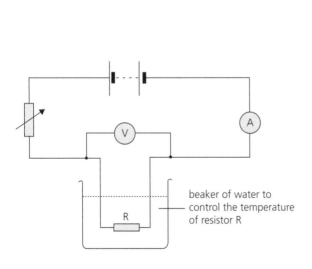

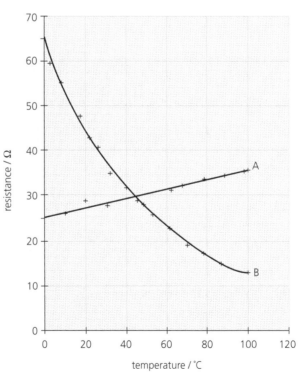

a) i) Which extra piece of apparatus is required to measure the temperature? *[1 mark]*

....

ii) Explain how to change the temperature of the water. *[1 mark]*

....

iii) What precautions should be taken to measure the correct temperature of the resistor? *[1 mark]*

....

iv) State how to use the graph to calculate the resistance of R at a known temperature. *[2 marks]*

....

....

b) Compare the way in which the resistances of the two resistors change as the temperature changes from 0 °C to 100 °C. *[4 marks]*

Resistor A: ....

....

....

Resistor B: ....

....

....

c) Which line, A or B, shows the resistance of the thermistor? Explain your answer. *[2 marks]*

....

....

d) The results show an anomaly. Identify the anomalous point on the graph and write down its co-ordinates. *[1 mark]*

....

e) i) What is the temperature when the resistors A and B have the same resistance? *[1 mark]*

....

ii) Use the graph to predict the resistance of resistor A at a temperature of 120 °C. *[2 marks]*

....

....

iii) Describe how the current in resistor A changes as the temperature increases.
Explain your answer. [1 mark]

....

(Total = 16 marks)

14 The instruction book for a hairdryer includes the safety warnings written below.

Explain why the user should **not** ignore these safety warnings:

a) 'Do not use this device near water contained in basins or bathtubs.' [3 marks]

....

....

....

....

b) 'Do not use this equipment if the cable is damaged or frayed.' [3 marks]

....

....

....

....

c) 'Do not unplug this appliance by pulling on the cord.' [3 marks]

....

....

....

....

(Total = 9 marks)

15 The diagram shows the connections inside a UK mains plug. It contains a fuse.

a) Explain why the wires are covered in plastic. [2 marks]

....

....

b) Describe how a fuse acts as a safety device. *[3 marks]*

....

....

....

....

c) State whether the fuse is placed in the circuit in the live wire, the neutral wire or the earth wire. *[1 mark]*

....

(Total = 6 marks)

16 Here is some information about an electric lawn mower:

- power 1100 W
- double insulated
- weight 11.5 kg
- cutting width 33 cm.

a) Explain what **double insulated** means. *[2 marks]*

....

....

b) The lawn mower must be used with a circuit breaker, even though it is double insulated. Explain why the circuit breaker is necessary. *[3 marks]*

....

....

....

....

c) State the equation linking power, current and voltage. *[1 mark]*

....

d) The lawn mower needs a new fuse. The fuses available are 3 A, 5 A and 13 A. State which fuse is most suitable for the lawn mower. Explain your answer. *[2 marks]*

....

....

....

(Total = 8 marks)

17 A student was given these instructions to draw the lighting circuit for a motorbike:

- The front light and the rear light are turned on together using one switch.

- The right and left indicator lights turn on using separate switches.

- If one light breaks, all other lights remain lit.

The student drew this circuit:

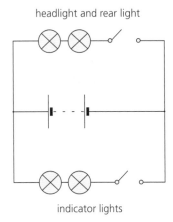

headlight and rear light

indicator lights

Identify **two** mistakes in the wiring in the circuit, which mean that the lights do not work as described. For each mistake, explain what could happen to the lights. [4 marks]

....

....

....

....

....

....

(Total = 4 marks)

18 The lights in a shed are wired using a switch and two lamps connected to the mains electricity supply in parallel.

Explain **two** reasons why a parallel circuit is suitable for a lighting circuit. [4 marks]

....

....

....

....

....

(Total = 4 marks)

19 A student investigates how the current through a bulb varies with the voltage across it.

The graph shows the results of the investigation.

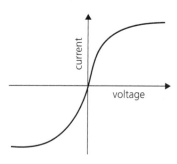

Describe how the shape of the graph changes as the voltage across the bulb changes. Explain these changes, linking your answer to changes in the resistance of the bulb. *[5 marks]*

....

....

....

....

....

....

....

(Total = 5 marks)

20 Electrostatic charge collects on an aeroplane when it is flying. Special precautions are needed before refuelling the aeroplane.

Explain why electrostatic charge on an aeroplane can be dangerous, and describe the precautions that should be taken during refuelling. *[5 marks]*

....

....

....

....

....

....

(Total = 5 marks)

21 The instruction book for a vacuum cleaner says:

'Always make sure the supply voltage used is the same as the voltage written on the vacuum cleaner.'

Explain why the vacuum cleaner may not work properly if the supply voltage is much higher or lower than the voltage it is designed for. [5 marks]

....

....

....

....

....

....

(Total = 5 marks)

22 Explain why a fuse is used in electric circuits, and describe how a fuse works. [5 marks]

....

....

....

....

....

....

(Total = 5 marks)

③ Waves

1 You are given a glass block, a protractor, paper, a pencil, a ruler and a ray box.

   Describe how to use this equipment to find the refractive index of the glass block.   *[5 marks]*

   ....

   ....

   ....

   ....

   ....

   ....

                                                                                                   (Total = 5 marks)

2 A student wrote this plan to measure the speed of sound in air.

   • Stand near a large wall.

   • Clap your hands and listen for the echo.

   • Clap in this pattern: clap – echo – clap – echo.

   • The speed of sound = $\dfrac{2 \times \text{distance to the wall}}{\text{time between a clap and its echo}}$

   Explain how you could improve the accuracy of this investigation.   *[4 marks]*

   ....

   ....

   ....

   ....

   ....

                                                                                                   (Total = 4 marks)

3 Design an experiment to measure the frequency of a sound wave. You can use some or all of
   this equipment: loudspeaker, a musical instrument (such as a recorder), microphone, stopclock,
   oscilloscope, ruler.   *[6 marks]*

   ....

   ....

   ....

   ....

....

....

(Total = 6 marks)

4 The diagram shows light shining through a glass block. Some light is refracted and some is reflected.

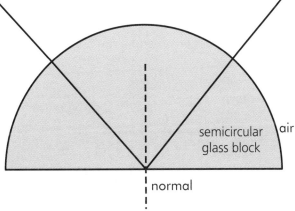

Which ray represents:

a) the incident ray in air   *[1 mark]*

....

b) the reflected ray in air   *[1 mark]*

....

c) the refracted ray in air?   *[1 mark]*

....

(Total = 3 marks)

5 The diagram shows total internal reflection in a semicircular glass block.

a) Use a protractor to measure the angle of incidence shown on the diagram, and write the angle below.   *[1 mark]*

....

b) State the equation linking refractive index and critical angle. *[1 mark]*

....

c) Calculate the refractive index of the prism. *[2 marks]*

....

....

....

(Total = 4 marks)

6 The image shows an oscilloscope screen. The trace is of a sound wave.

The time base setting on the oscilloscope is 10 ms per square.

Calculate the frequency of the sound wave. *[4 marks]*

....

....

....

....

....

(Total = 4 marks)

7 Describe how a student could use a glass prism and a ray box to investigate how light is refracted by a triangular prism. *[5 marks]*

....

....

....

....

....

....

(Total = 5 marks)

8 The time period of the sound from a car horn is 0.0025 s. The speed of sound is 340 m/s.

What is the wavelength of this sound wave? Circle the correct answer. *[1 mark]*

A 0.85 m

B 400 m

C 160 km

D 340 m

(Total = 1 mark)

9 A pulse of ultrasound is transmitted downwards from a ship at sea. The ultrasound pulse reflects off the sea bed, 1.8 km below, and the reflection is detected on the ship. The ultrasound wave travels at 1500 m/s in water.

How long does it take for a detector on the ship to detect the reflected pulse? Circle the correct answer. *[1 mark]*

A 0.0012 s

B 0.0024 s

C 2.4 s

D 1.2 s

(Total = 1 mark)

10 The diagram shows a transverse wave on a piece of thick string. The wave is produced by a person holding the end, A, of the string and moving it up and down. The end A moves up and down five times in 2 s.

The wave moves to the right, past a grid, which is marked with 10 cm × 10 cm squares.

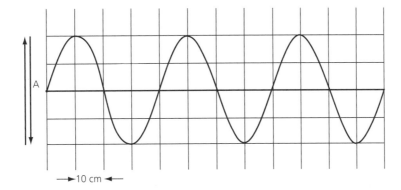

a) Explain what is meant by a **transverse wave**. *[2 marks]*

....

....

b) Use the diagram to calculate the wave's:

i) amplitude *[1 mark]*

....

Photocopying prohibited

ii) wavelength. *[2 marks]*

....

....

c) Calculate the wave's:

i) frequency in Hz *[2 marks]*

....

....

ii) time period. *[2 marks]*

....

....

d) Calculate the speed of the wave. *[3 marks]*

....

....

....

....

e) Make a sketch on the original diagram to show a wave, made by the same thick string, which has half the amplitude and twice the frequency of the wave already shown in the diagram. *[2 marks]*

(Total = 14 marks)

11 A student carries out an experiment to find the refractive index of a glass block. Here are some measurements.

Angle of incidence: 40°

Angle of refraction: 26°

What is the refractive index for this glass block? Circle the correct answer. *[1 mark]*

A 1.54

B 0.65

C 0.977

D 1.47

(Total = 1 mark)

12 When waves pass through gaps, the waves diffract.

a) Explain the meaning of the term **diffraction**. *[1 mark]*

....

b) The two diagrams below show water waves in ripple tanks approaching two different gaps in a barrier.

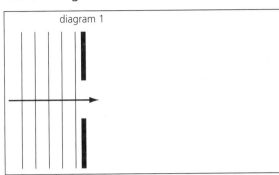 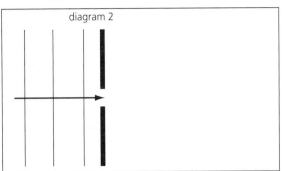

Complete the two diagrams to show what happens to each group of waves as they pass through the barrier. [4 marks]

c) The mouse-eared bat produces sound waves with a frequency of 50 kHz. The speed of these sound
waves in air is 330 m/s.

i) State the equation that links wave speed, frequency and wavelength. [1 mark]

....

ii) Calculate the wavelength of the waves. [3 marks]

....

....

....

iii) Explain why the sound waves help the bats find moths to eat. [2 marks]

....

....

d) The wavelengths of sound produced by people talking are usually in the range 0.5 m to 1.5 m.

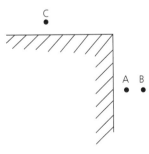

Explain why the person standing at C can hear the people talking at A and B, even though she cannot see them. [2 marks]

....

....

....

(Total = 13 marks)

Photocopying prohibited

13 A girl looks at the reflection of a coin using a plane mirror.

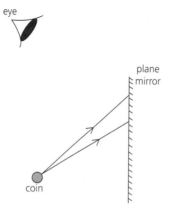

a) Complete the ray diagram to show how the rays of light enter her eye. [3 marks]

....

....

....

b) i) Mark an **X** on the diagram to show where the girl sees the image. [2 marks]

   ii) What type of image does she see? Is it real or virtual? Explain your answer. [2 marks]

....

....

(Total = 7 marks)

14 The diagram shows how a semicircular glass block can be used to determine the critical angle for this type of glass. A ray of light (AB) is directed towards the midpoint, B, of the glass block. At the critical angle of 40°, some light is reflected along BD, and some of the light is refracted along the surface of the block – this is the ray BC.

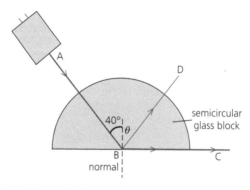

a) In the diagram, some light is reflected along the line BD. State the angle, θ, between the normal and BD. [1 mark]

....

b) Show, by drawing on the diagram, what happens when the ray box is moved so that:

   i) the angle between ray AB and the normal is greater than 40° [2 marks]

   ii) the angle between the ray AB and the normal is less than 40°. [3 marks]

c) Calculate the refractive index of the glass. *[3 marks]*

....

....

....

....

(Total = 9 marks)

15 People to the south of a large hill receive some radio stations very clearly on their radios.

The reception of other radio stations is so bad that the people cannot hear them. The radio transmitter is to the north of the hill.

The table shows the frequency and wavelength of different radio stations.

Radio station	Frequency	Wavelength / m
Radio 1	98 MHz	3.0
Radio 2	89 MHz	3.4
Radio 4	198 kHz	1515.0
local radio	104 MHz	2.9

Use your ideas about diffraction to explain which radio stations are received most clearly on the south side of the hill. *[4 marks]*

....

....

....

....

....

....

(Total = 4 marks)

16 In some countries, including the UK, it is illegal for people under 18 to use a sunbed in a salon. The number of people with skin cancer has risen over the last 40 years. One cause of skin cancer is exposure to ultraviolet radiation from sunbeds and sunlight. This graph shows new cases of skin cancer per 100 000 people in the UK for the period 1975–2010.

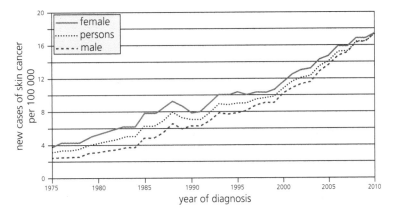

Use your ideas about electromagnetic radiation to discuss how this law could affect the number of skin cancer cases in the UK in future. *[4 marks]*

....

....

....

....

....

....

(Total = 4 marks)

17 A satellite dish on the side of a house receives microwave signals carrying satellite television broadcasts. The wavelength of microwaves is about 3 cm.

Use ideas about diffraction to explain why the diameter of this satellite dish is about 50 cm. *[4 marks]*

....

....

....

....

....

....

(Total = 4 marks)

18 *Silent Whistles* are made for dog owners to use. The manufacturer claims:

- dogs and young children can hear the whistle

- adults cannot hear the whistle

- dogs can hear the whistle through background noise from traffic.

The tables show some typical frequencies of sound sources and frequencies that can be heard by different animals.

Animal	Frequency range of hearing / Hz
human (child)	20–20 000
human (adult)	20–14 000
dog	40–60 000

Source of sound	Frequency range of sound produced / Hz
dog whistle	16 000–22 000
traffic noise	500–1800

Discuss whether the *Silent Whistle* manufacturer's claims are correct. [4 marks]

....

....

....

....

....

....

(Total = 4 marks)

19 Microwave ovens have metal shielding inside. There is also metal mesh in the glass doors
to stop microwaves passing through. Instructions for the microwave oven state the microwave
oven must not be used if the door is damaged.

Staff working in a café who use the microwave oven are told not to look through the door when it
is turned on because this could be dangerous.

Explain whether this precaution is necessary. [4 marks]

....

....

....

....

....

(Total = 4 marks)

20 Explain how electromagnetic radiation is used in these applications:

a) fluorescent lights [3 marks]

....

....

....

b) night vision equipment. [3 marks]

....

....

....

(Total = 6 marks)

21 Food irradiation is used to sterilise some types of foods that are perishable. It reduces the number of cases of food poisoning caused by bacteria.

Explain why X-rays and gamma rays are used to sterilise foods. [4 marks]

....

....

....

....

....

(Total = 4 marks)

22 Microwave radiation is an example of electromagnetic radiation.

State **two** uses for microwave radiation, and explain how the properties of microwave radiation make it suitable for each use. [4 marks]

....

....

....

....

....

....

(Total = 4 marks)

23 A teacher shows her class how a longitudinal wave and a transverse wave transfer energy through a long slinky spring.

Describe how the teacher can demonstrate longitudinal waves and transverse waves using the slinky spring, and compare the two different types of wave. [5 marks]

....

....

....

....

....

(Total = 5 marks)

24 A company uses ultrasound to detect cracks inside metal blocks.

An ultrasound probe sends ultrasound pulses into the metal block. A receiver detects when the reflected pulses return. These are displayed on a screen. The ultrasound trace from one metal block is shown in the diagram.

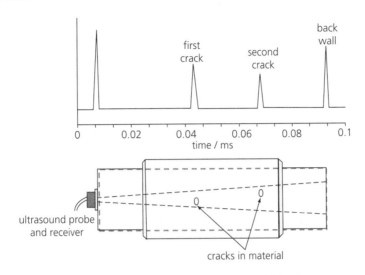

Explain how ultrasound is used to detect cracks inside the metal block. State **one** advantage for the company of using ultrasound to test the metal blocks. *[5 marks]*

....

....

....

....

....

....

(Total = 5 marks)

25 A doctor uses an endoscope to see inside a patient's knee. The endoscope uses two bundles of optical fibres, which pass through a small cut in the patient's knee.

Explain how the endoscope helps the doctor see images from inside the patient, and explain **one** advantage of using optical fibres in an endoscope. *[5 marks]*

....

....

....

....

....

....

(Total = 5 marks)

Energy resources and energy transfers

4

1 A commercial hovercraft service runs in the UK between Portsmouth and the Isle of Wight. The hovercraft has a mass of 60 000 kg and a maximum speed of 13 m/s.

The hovercraft approaches the beach and slows down.

a) State the equation linking kinetic energy, mass and velocity. *[1 mark]*

....

b) Calculate the change in kinetic energy, in J, when the hovercraft slows down from 13 m/s to 10 m/s. *[3 marks]*

....

....

....

(Total = 4 marks)

2 The diagram shows the design of an experiment to investigate the efficiency of an electric motor.

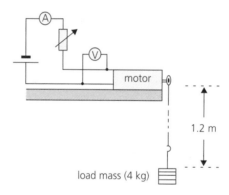

a) A load of mass 4 kg is lifted through a height of 1.2 m in 2.4 s.

i) State the equation that links gravitational potential energy to mass, g and height. *[1 mark]*

....

ii) Calculate the increase in the gravitational potential energy of the load, in J.
g = 10 N/kg *[2 mark]*

....

....

b) When the load is being lifted, the current in the circuit is 5.0 A, and the voltage across the motor 16.0 V. The time taken is 2.4 s.

i) State the equation that links energy transferred, voltage, current and time.   [1 mark]

....

ii) Calculate the energy transferred to the motor in the time it takes to lift the load through a distance of 1.2 m.   [2 marks]

....

....

iii) Calculate the efficiency of the motor while it lifts the load.   [2 marks]

....

....

c) Explain where energy is wasted in the process of lifting the load.   [2 marks]

....

....

(Total = 10 marks)

3   The Sankey diagrams compare energy transfers in a filament lamp and an energy-saving lamp.

Sankey diagram for a filament lamp

Sankey diagram for an energy-saving lamp

60 J electrical energy

heat energy

3 J light energy

electrical energy

45 J heat energy

15 J light energy

a) What is the total output energy of the filament bulb? Circle the correct answer.   [1 mark]

A 60 J

B 3 J

C 57 J

D 120 J

b) Calculate the efficiency of the filament lamp.   [3 marks]

....

....

c) A student decides to replace a 60 W filament bulb. Energy-saving bulbs of power 7 W, 12 W, 18 W or 23 W are available.

Explain which energy-saving bulb will give the same brightness as the 60 W filament bulb.   *[3 marks]*

....

....

....

....

(Total = 7 marks)

4 Why are hydroelectric dams a useful way to generate renewable electricity? Circle the correct answer.   *[1 mark]*

A Electrical energy can be stored in the turbine until it is needed.

B Water can be stored in the upper reservoir until it is needed.

C Constant water pressure on the dam generates constant electricity.

D Hydroelectricity can be used anywhere there is a river.

(Total = 1 mark)

5 Meteors are formed from small particles of dust that orbit the Sun. Sometimes meteors enter the Earth's atmosphere and cause a bright streak of light.

a) A meteor with a mass of 0.0 003 kg enters the Earth's atmosphere travelling at a velocity of 18 000 m/s.

Calculate the meteor's kinetic energy. State the unit of energy.   *[3 marks]*

....

....

b) Explain what happens to the meteor's kinetic energy as it passes through the atmosphere, and use these ideas to suggest why very few meteors hit the ground.   *[4 marks]*

....

....

....

....

(Total = 7 marks)

6 Explain how the different features of a sleeping bag reduce heat losses for a person sleeping in it. [4 marks]

....

....

....

....

(Total = 4 marks)

7 A firework rocket is to be used as part of a firework display in the sky. Describe the energy transfers that happen when the firework is lit and launches itself into the firework display. Your answer should include:

- the energy transfers involved
- where these transfers take place
- how energy is conserved. [6 marks]

....

....

....

....

....

....

(Total = 6 marks)

8 Solar tiles are installed on the roofs of buildings. They use thermal (heat) energy from the Sun to heat water used in the home. Explain the features of a solar tile that make it effective at heating water. [6 marks]

cold water in
glass cover
warm water out
water flows through metal tubes
back surface for insulation painted black

....

....

....

....

....

....

(Total = 6 marks)

9 A new power station is needed to supply 10 000 homes with electricity. The houses are in a city with good road and rail transport links to the site. The developer must choose between a biomass power station or a combined heat and power station.

Biomass power station	Combined heat and power station
generates electricity	generates electricity and heat for homes
fuel is wood pulp transported by lorry	fuel is natural gas transported through a pipeline
the wood pulp comes from a factory 30 km away	waste heat from power station is transported in pipes to local homes to provide heating

a) Explain **one** advantage and **one** disadvantage of each power station. [4 marks]

....

....

....

....

b) Suggest, with a reason, which power station is more suitable for the site. [3 marks]

....

....

....

....

(Total = 7 marks)

10 A firm sells LED lamps for electrical lighting in the home. These lamps cost ten times more than halogen lamps. Its advert says:

- 'Cool, super-bright, 90% energy-saving LED lamp.'

- 'A 9 W LED lamp gives the same light as a 50 W halogen lamp.'

- 'LED lamps last for 50 000 hours (up to 15 years normal use).'

- '1 LED lamp lasts as long as 50 halogen lamps.'

- 'The payback time is 10 months.'

Explain the advantages to the environment if LED lamps are used instead of halogen lamps in many homes. Your answer should consider the effects of producing, using and disposing of the bulbs. [4 marks]

....

....

....

....

....

(Total = 4 marks)

11 The diagram shows the structure of a double-glazing unit. Explain how double glazing reduces thermal (heat) losses from homes. [4 marks]

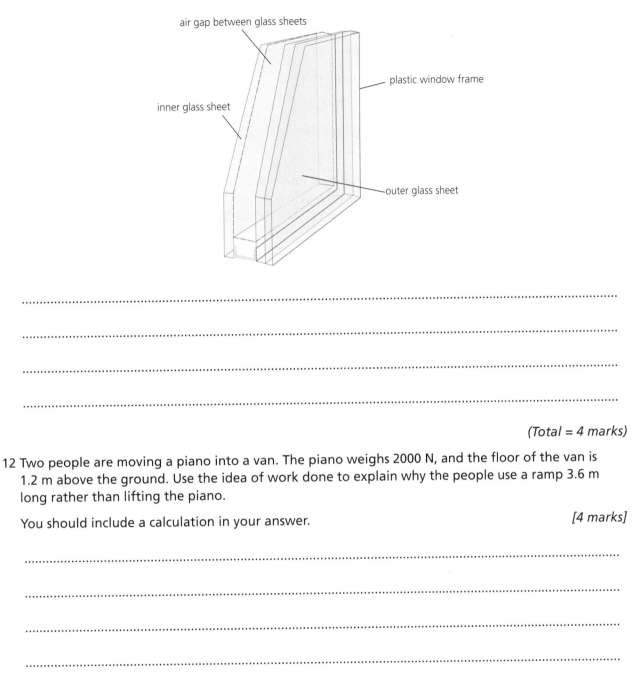

air gap between glass sheets

plastic window frame

inner glass sheet

outer glass sheet

....

....

....

....

(Total = 4 marks)

12 Two people are moving a piano into a van. The piano weighs 2000 N, and the floor of the van is 1.2 m above the ground. Use the idea of work done to explain why the people use a ramp 3.6 m long rather than lifting the piano.

You should include a calculation in your answer. [4 marks]

....

....

....

....

....

(Total = 4 marks)

13 A person makes a bungee jump from a bridge. The bungee rope is elasticated.

What is the main energy change taking place when the person first jumps from the bridge until they first come to a stop at the bottom of their jump? Circle the correct answer.   [1 mark]

A elastic energy store → kinetic energy store

B gravitational energy store → kinetic energy store

C gravitational energy store → elastic energy store

D elastic energy store → gravitational energy store

(Total = 1 mark)

14 Allstown is a town on the coast that needs a new power station. Allstown has a busy port and good road and rail links, which are shown on the map below. The land around Allstown is flat. The power station built must supply at least 30 MW of power, enough for 20 000 homes, at all times.

These projects have been suggested:

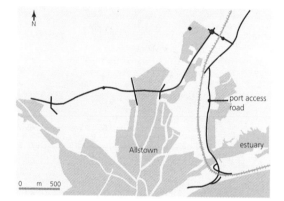

- An offshore wind farm with 12 turbines. Each turbine generates 3 MW in suitable weather conditions. In future, more turbines may be added.

- A biomass plant using wood pulp and waste from local businesses. It generates 35 MW at all times.

- A combined heat and power station that uses natural gas. This generates 32 MW of electricity and provides heating directly to 5000 homes from waste heat produced by the power station.

Use your ideas about the electricity production to recommend the most suitable power station for Allstown. You should also explain why you do not recommend the other projects.   [5 marks]

....

....

....

....

....

....

....

(Total = 5 marks)

15 Here is some information about an electric mobility scooter, RangeMax.

- The RangeMax scooter has a top speed of 15 km/h.

- The RangeMax weighs 40 kg.

- It can be used by people weighing up to 80 kg.

- It uses a rechargeable battery, powering the scooter for a maximum range of 10 km.

Use your ideas about energy transfers to explain why:

- the scooter cannot travel at its top speed of 15 km/h uphill

- the range of the scooter is less if it is used in hilly areas.

You may include calculations in your answer.   *[5 marks]*

....

....

....

....

....

....

....

(Total = 5 marks)

5 Solids, liquids and gases

1 Thermometers are used in a science lesson to measure how the temperature of a beaker of water changes over time. All thermometers are at room temperature. The thermometers do not all show the same temperature reading.

a) Explain **two** ways students can allow for different temperature readings from different thermometers when they carry out the experiment. [4 marks]

....

....

....

....

b) The experiment is to measure how the temperature of a beaker of hot water changes over time.

The students work in pairs for this experiment. Explain **one** benefit of working in pairs for this experiment. [3 marks]

....

....

....

....

(Total = 7 marks)

2 Robert Brown was a scientist who watched the motion of pollen grains in water carefully, using a microscope. The pollen grains were constantly moving. This movement was named Brownian motion.

A teacher set up an experiment for students to see Brownian motion.

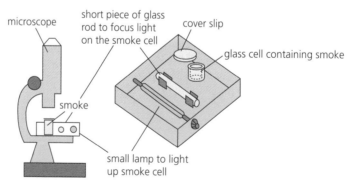

a) The glass cell contains smoke. Explain how smoke particles are used to show Brownian motion. [2 marks]

....

....

b) Light from a lamp shines on to the glass cell. Explain why the lamp is used in the experiment. *[2 marks]*

....

....

c) The picture shows a drawing the students made showing the path of some smoke particles.

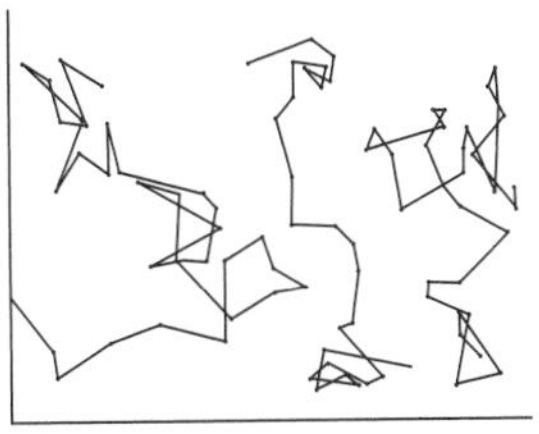

Explain why the results of this experiment led scientists to believe that atoms exist. *[5 marks]*

....

....

....

....

....

....

(Total = 9 marks)

3 A student uses a measuring cylinder, 30 cm ruler and a top-pan balance to measure the density of a cube of brass

a) Explain how the student should use the equipment to make her measurements as accurate as possible. *[4 marks]*

....

....

....

....

....

b) Each side of the brass cube measures 2.0 cm. The density of brass is 8400 kg/m³. What is the mass of the brass cube? Circle the correct answer. *[1 mark]*

A 67 200 kg

B 0.0672 kg

C 1050 kg

D 33.6 kg

(Total = 5 marks)

4 A bottle is standing on a table.

a) Describe an experiment to calculate the pressure the bottle exerts on the table.   *[4 marks]*

....

....

....

....

....

b) The bottle's mass is 0.2 kg. The diameter of its base is 10 cm. How much pressure does the bottle exert on the table? Circle the correct answer.   *[1 mark]*

A 64 Pa

B 25 Pa

C 6.4 Pa

D 255 Pa

(Total = 5 marks)

5 The Kelvin scale and the Celsius scale are used to measure temperature.

a) Describe **two** similarities and **two** differences between the temperature scales.   *[4 marks]*

....

....

....

....

....

b) State how to convert a reading in Kelvin to a reading in degrees Celsius.   *[1 mark]*

....

....

(Total = 5 marks)

6 A swimmer takes an inflated balloon underwater in a swimming pool.

Compare the size and shape of the balloon when it is held just under the surface of the water, and when it is held at the bottom of the swimming pool.   *[4 marks]*

....

....

....

....

....

....

(Total = 4 marks)

7 A games teacher inflates two identical netballs to the same pressure, then takes one netball outside and leaves the other netball in the sports building.

The table shows the temperature readings in the building and outside.

Place	Temperature reading / °C
outside the sports building	2
inside the sports building	20

An hour later, a student checks the pressure in both netballs before the game starts. The pressure in one netball is higher than the pressure in the other netball. Use your ideas about air pressure and molecules to explain:

a) why the pressure inside the balls increases as they are pumped up   *[3 marks]*

....

....

....

b) why the pressure of the two balls is not the same an hour later after they are left
   in different places.   *[3 marks]*

....

....

....

(Total = 6 marks)

8 A student weighing 700 N stands up. The area of each shoe in contact with the floor is 0.02 m². Calculate the pressure exerted on the floor by the student when the student:

a) stands on one foot   *[2 marks]*

....

....

b) stands on both feet.   *[2 marks]*

....

....

(Total = 4 marks)

6 Magnetism and electromagnetism

1 A student wrote a plan to compare the strength of the magnetic field around two different types of magnet.

- I will put one magnet on the bench and collect a compass.
- I will see if the compass needle changes direction when it is by the magnet.
- Then I will put another magnet on the bench by the first one.
- I will get another compass and see if the compass needle changes direction.
- If it does the second magnet is stronger than the first one.

Identify **three** faults in the plan, and correct each one. *[6 marks]*

....

....

....

....

....

....

(Total = 6 marks)

2 A student can either use plotting compasses or iron filings to investigate the magnetic field around a coil of wire.

a) Circle **one** advantage of using several plotting compasses instead of iron filings. *[1 mark]*

A Plotting compasses show the position of the field lines.

B Plotting compasses are not affected by bar magnets.

C Plotting compasses always point north.

D Plotting compasses show the direction of the magnetic field.

b) Describe how the student can show that the magnetic field lines around the coil of wire change when the current is turned on and off. *[3 marks]*

....

....

....

(Total = 4 marks)

3 The diagram shows an experiment to investigate electromagnetic forces.

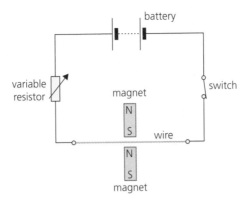

a) A wire passes between the two magnets. Describe what happens when the student switches the circuit on.   [3 marks]

....

....

....

b) Explain how the variable resistor can be used to increase the size of the force on the wire.   [2 marks]

....

....

c) The student investigates the effect on the size of the magnetic force when the wire is made from different metals. Which two variables should the student control when choosing the wires to test? Circle the correct answer.   [1 mark]

A wire material and wire length

B wire length and connecting wires

C wire material and wire thickness

D wire length and wire thickness

(Total = 6 marks)

4 The generator in a power station produces electricity at a voltage of 25 000 V.

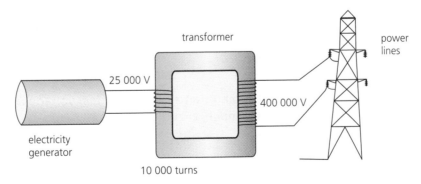

A transformer steps this voltage up to 400 000 V, for the electricity to be transmitted on the country's power lines.

a) i)   Use the information in the diagram to calculate the number of turns needed in the secondary coil of the step-up transformer.

[3 marks]

....

....

....

ii) The current in the primary coil is 800 A. Calculate the current that is carried by the power lines.

[3 marks]

....

....

....

b) Explain why very high voltages are used to transmit currents over long distances.

[3 marks]

....

....

....

(Total = 9 marks)

5   The diagram shows a laboratory demonstration of how a transformer may be used to produce very large currents. In this example, a large current is being used to melt a nail.

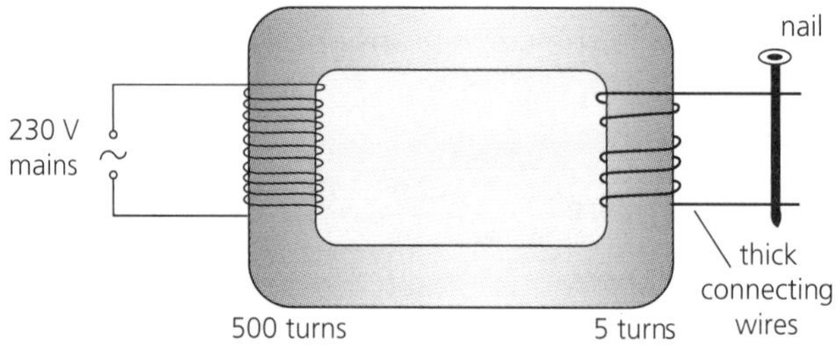

a) Calculate the voltage cross the nail. Circle the correct answer.

[1 mark]

A   230 V a.c.

B   230 V d.c.

C   2.3 V a.c.

D   11.5 V a.c.

b) The nail has a resistance of 0.02 Ω. Assuming that all the secondary voltage is across the nail, calculate the current in the secondary circuit.

[2 marks]

....

....

....

c) Explain why the secondary coil is made from very thick wire. *[1 mark]*

....

....

d) Show that the power generated in the secondary circuit is 265 W. *[2 marks]*

....

....

....

e) Calculate the current in the primary circuit, assuming the transformer is 100% efficient. *[2 marks]*

....

....

....

f) The nail melts after 15 000 J of energy are transferred to it. Calculate how long it takes the nail to melt. *[3 marks]*

....

....

....

(Total = 11 marks)

6   Seismic waves from earthquakes are detected using instruments called seismometers. The diagram shows a simple seismometer.

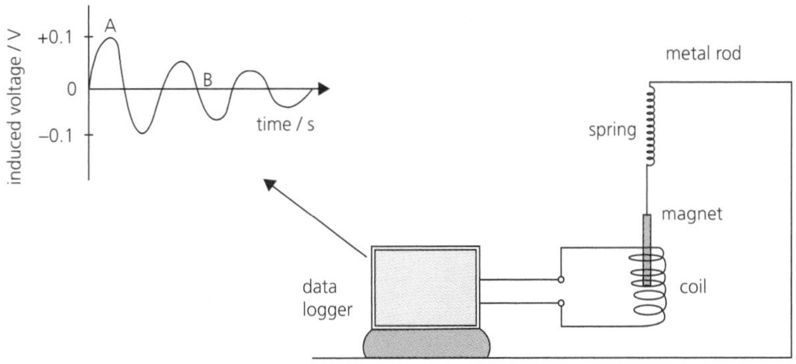

The seismometer consists of a bar magnet suspended from a spring. The magnet hangs from a spring, which is attached to a metal rod, which transmits vibrations from earthquakes. The magnet moves up and down inside a coil when there is an earthquake. A data logger records the voltage across the coil.

a) i)   Explain why a voltage is induced in the coil. *[1 mark]*

....

....

ii) Explain why this voltage is alternating. *[1 mark]*

....

....

b) The graph in the diagram shows the form of the induced voltage when there is an earthquake.

i) Describe the motion of the magnet when the induced voltage has its greatest value, at the point labelled A. *[2 marks]*

....

....

....

ii) Describe the motion of the magnet when the induced voltage is zero at the point labelled B. *[2 marks]*

....

....

c) Explain **two** ways in which you could make the seismometer more sensitive. *[2 marks]*

....

....

....

(Total = 8 marks)

7 a) Use the left-hand rule to predict what happens in the following examples. The arrows in each diagram show the direction of the current.

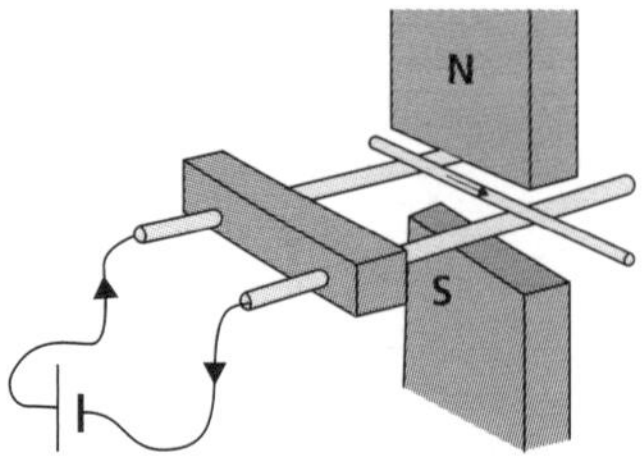

diagram A

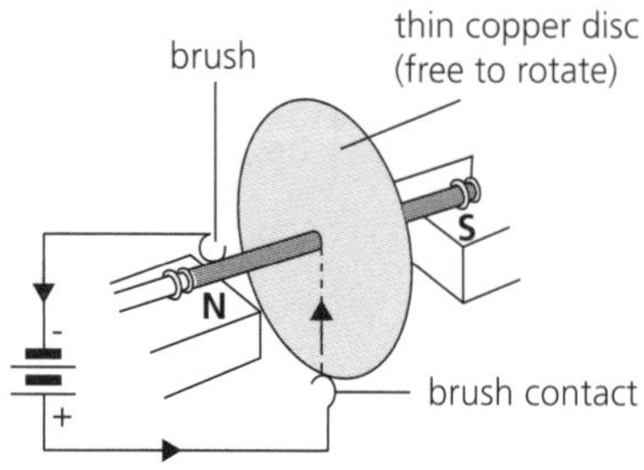

diagram B

i) In which direction does the rod move in diagram A? *[1 mark]*

....

ii) Explain why the disc rotates in diagram B. *[2 marks]*

....

....

iii) Does the disc rotate in a clockwise or an anticlockwise direction?
Explain your answer. *[2 marks]*

....

....

b) The diagram shows a model electric motor. In this diagram, the current flows round the motor coil, in the direction A to B to C to D.

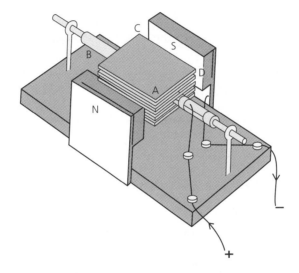

i) What is the direction of the force on side AB? *[1 mark]*

....

ii) What is the direction of the force on side CD? *[1 mark]*

....

iii) In which direction does the coil rotate, when viewed from the right? *[1 mark]*

....

(Total = 8 marks)

8   The diagram shows a bicycle dynamo. When a bicycle wheel moves, it turns the driving wheel of the dynamo. The driving wheel is attached to a permanent magnet, which rotates inside a fixed coil.

The output terminals of the dynamo are connected to a data logger, and the output voltage is displayed on an oscilloscope. The oscilloscope trace below shows the output voltage when the drive wheel is rotating ten times each second:

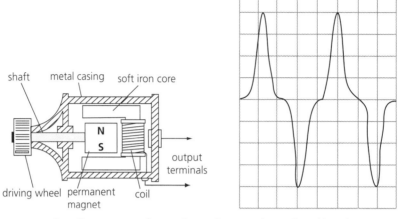

a) Add a second trace to the diagram to show the voltage when the wheel rotates five times each second.   *[2 marks]*

b) Explain why a bicycle lamp attached to the dynamo does not light when the bicycle is stationary.   *[2 marks]*

....

....

(Total = 4 marks)

9 The diagram shows the structure of a loudspeaker.

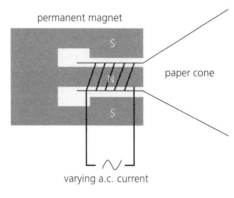

Explain, in as much detail as possible, how the loudspeaker produces a sound when an a.c. current flows.   *[5 marks]*

....

....

....

....

....

(Total = 5 marks)

10 The diagram shows the structure of an electric bell. Explain in detail why the bell rings
   when the switch is pressed.   *[6 marks]*

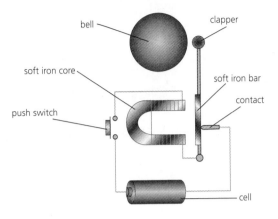

....

....

....

....

....

....

(Total = 6 marks)

11 The diagram shows how parts of the UK National Grid are linked together.

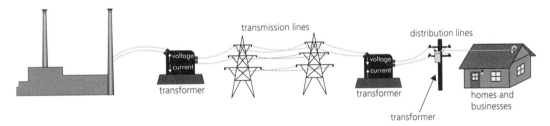

Describe how transformers are used in the National Grid to transfer electrical energy
efficiently from the power station to homes and businesses.   *[6 marks]*

....

....

....

....

....

....

(Total = 6 marks)

12 The diagram shows the structure of a transformer.

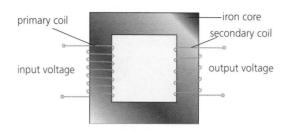

This transformer is designed to decrease the voltage of an a.c. supply.
Explain how it works.

[5 marks]

....

....

....

....

....

(Total = 5 marks)

7 Radioactivity and particles

1 Plutonium-241 is a radioisotope that eventually decays to the stable isotope bismuth-83.

All the steps in this sequence of decays are shown on the diagram. The symbols of the elements are included in the diagram. The table links the chemical symbols for the elements to their names.

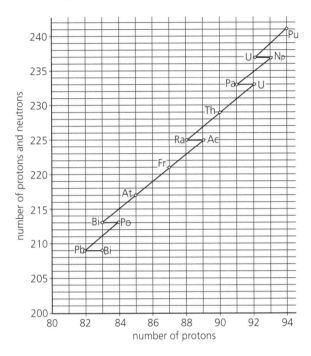

Element	Symbol
plutonium	Pu
neptunium	Np
uranium	U
protactinium	Pa
thorium	Th
actinium	Ac
radium	Ra
francium	Fr
astatine	At
polonium	Po
bismuth	Bi
lead	Pb

a) i) The nuclei of elements can be described in the form: $^{241}_{94}$Pu.

What particles are found in a nucleus of plutonium-241? Circle the correct answer.   *[1 mark]*

A 94 protons, 241 neutrons

B 94 electrons, 147 neutrons

C 94 neutrons, 147 protons

D 94 protons, 147 neutrons

ii) Use the information in the diagram to describe the isotopes of thorium and francium in the form shown above.   *[2 marks]*

....

....

....

b) i) Which particle is emitted in this decay of polonium to lead?   *[1 mark]*

....

ii) Construct an equation to show the decay of $^{213}_{84}$Po. *[2 marks]*

....

....

c) i) Which particle is emitted in this decay of actinium to francium?
Circle the correct answer. *[1 mark]*

A beta particle

B alpha particle

C beta particle and alpha particle

D neutron

ii) Construct an equation to show the decay of $^{255}_{88}$Ra. *[2 marks]*

....

....

d) The element uranium appears twice in this sequence of decays.

Explain the difference between these two types of uranium. *[2 marks]*

....

....

....

e) Calculate the total numbers of alpha particles and beta particles emitted in this sequence of decays. *[1 mark]*

....

....

(Total = 12 marks)

2 Wood contains a lot of carbon, most of which is the isotope carbon-12 or $^{12}_{6}$C.

Wood also contains a small fraction of the radioactive isotope carbon-14 or $^{14}_{6}$C.

The half-life of carbon-14 is 5700 years. Carbon-14 decays to nitrogen-14.

While a tree is alive, it has a constant fraction of carbon-14. But, after the tree dies, this fraction decreases because of radioactive decay. Measurement of this fraction allows us to work out the age of very old samples of wood.

a) Calculate the number of neutrons in a nucleus of:

i) carbon-12 *[1 mark]*

....

ii) carbon-14. *[1 mark]*

....

b) What particle is emitted by a carbon-14 nucleus when it decays?
   Circle the correct answer. *[1 mark]*

   A neutron

   B alpha particle

   C beta particle

   D beta particle and alpha particle

c) A scientist compares two samples of wood: one from a recently felled tree, the other from an older piece of wood. After corrections to account for the background count and mass of the samples, the scientist finds the count rate from the old wood is four times lower than the count rate from the new wood.

   i) Calculate the age of the old wood. *[2 marks]*

   ....

   ....

   ....

   ii) Explain why it is necessary to make a correction to account for the background count. *[2 marks]*

   ....

   ....

   ....

   iii) Explain **one** precaution you must make to ensure there is a fair comparison
        between the old and new wood. *[1 mark]*

   ....

   ....

   (Total = 8 marks)

3 A teacher demonstrates how to measure the half-life of radon (a gas that emits alpha particles) using this apparatus:

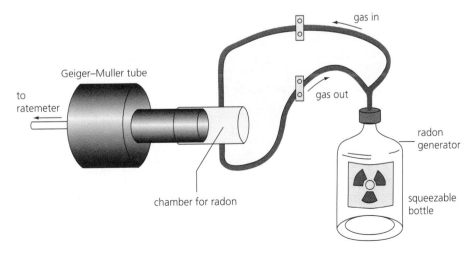

a) Describe, explaining which part of the body would be affected most, how radon gas can be harmful if it leaks out of the container. *[3 marks]*

....

....

....

....

....

b) The teacher obtained these results in her experiment.

Count rate / Bq	320	283	250	221	195	172	152	135	119	105
Time / s	0	10	20	30	40	50	60	70	80	90

Plot a graph of *count rate* (*y*-axis) against *time* (*x*-axis). *[4 marks]*

c) Use the graph to calculate the half-life of radon. *[2 marks]*

....

....

d) Calculate how long it takes for the count rate to fall below 10 Bq. *[2 marks]*

....

....

....

(Total = 11 marks)

4 Tritium is an isotope of hydrogen. Tritium has a mass number of 3 and an atomic number of 1. The most common isotope of hydrogen has a mass number of 1 and a proton number of 1.

a) The table lists the sub-atomic particles found in the nucleus and orbiting the nucleus. Circle the correct answer, choosing from A to D. *[1 mark]*

	Sub-atomic particle found in the nucleus	Sub-atomic particle orbiting the nucleus
A	electron, neutron	proton
B	electron, proton	neutron
C	neutron, proton	electron
D	neutron, electron	proton

b) Tritium decays by emitting a beta particle. The tritium nucleus turns into a helium-3 nucleus.

i) Draw a diagram to show the helium-3 nucleus. Label its particles. *[3 marks]*

ii) Construct an equation to describe the beta decay of $^3_1$H. *[2 marks]*

....

(Total = 6 marks)

5 An oil company is testing oil flow between two underground oil wells. It takes 2 to 3 months for oil to flow between the wells. A radioactive tracer is injected into oil in one well. Radioactivity levels in the other well are sampled and tested regularly for 1 year.

The table shows the tracers available:

Isotope	Half-life	Decay type
silver-110	8 months	beta
cobalt-60	5 years	beta and gamma
antimony-124	2 months	beta and gamma
phosphorus-32	14 days	beta

Which is the most suitable isotope to use? Explain your answer. *[4 marks]*

....

....

....

....

....

(Total = 4 marks)

6   A student's revision notes included these statements:

*'Gamma radiation is the most dangerous radiation because it can penetrate everything.
It is used in medical tracers.'*

Explain how the student's notes could be improved.   *[3 marks]*

....

....

....

....

....

(Total = 3 marks)

7   A teacher is demonstrating radioactive sources.

Before the lesson, the radioactive sources were stored in a lead-lined wooden box.

The teacher wears gloves and uses tongs to take the sample out of the box.

a) Describe **two** more precautions the teacher should take when demonstrating
    the samples.   *[2 marks]*

....

....

....

b) Explain which of the teacher's precautions is most important when showing students
    radioactive samples.   *[3 marks]*

....

....

....

....

(Total = 5 marks)

8   People working with radioactive sources must wear a film badge to monitor their exposure to
    radiation.

The badges are analysed every 2 months to calculate how much radiation the badges have been
exposed to. Results take 1 month to come back.

Explain how the film badge helps people work more safely with radioactive sources in hospitals, and
why the badges should not be left in the room with the sources when the person is not in
the same room.   *[4 marks]*

....

....

....

....

....

....

(Total = 4 marks)

9   A doctor tests the blood flow through a patient's heart by injecting a radioactive tracer. There is a radiation detector outside the patient's body to track the blood flow. The doctor uses this detector for half an hour immediately after injecting the tracer, and again 4 hours later.

The table shows the tracers available.

Isotope	Half-life	Decay type
yttrium-90	64 hours	beta
thallium-201	73 hours	gamma
magnesium-27	9.5 minutes	beta and gamma
chromium-51	27 days	alpha

Which is the most suitable isotope to use as the tracer? Explain your answer.   *[4 marks]*

....

....

....

....

....

....

(Total = 4 marks)

10 Nuclear waste contains different radioactive isotopes. Some of these isotopes have a half-life of days or weeks. Other isotopes have half-lives of more than a thousand years.

Highly radioactive waste is treated to make it safer. It is vitrified (changed into a glass-like material that does not dissolve). The waste is then sealed in a steel cylinder. The cylinders are stored in secure underground chambers.

a) Explain how treating and storing radioactive waste helps to reduce the risks of exposure to radiation.   *[2 marks]*

....

....

....

b) Explain why the waste becomes less dangerous after 100 years. *[2 marks]*

....

....

....

(Total = 4 marks)

11 Smoke detectors installed in homes include a source of alpha radiation. The alpha source is sealed in a plastic container inside the smoke detector, which is fixed to the ceiling. Explain whether the plastic container is enough to protect people from the alpha source inside the smoke detector. *[2 marks]*

....

....

....

(Total = 2 marks)

12 Cosmic radiation comes from outer space and includes gamma rays and X-rays. The Earth's atmosphere absorbs cosmic radiation, so the radiation is less intense on the ground but more intense for people travelling by aeroplane.

Here is an extract from a leaflet for pilots and aircrew:

- 'Aeroplanes flying above 15 km use a radiation detector. If high levels of cosmic radiation are detected, the aeroplane must fly lower.'

- 'Pilots cannot spend more than 500 hours per year flying higher than 15 km.'

Explain why cosmic radiation can be dangerous for pilots and how the risk can be reduced. *[5 marks]*

....

....

....

....

....

....

....

(Total = 5 marks)

13 Two scientists, Geiger and Marsden, carried out an experiment in 1908. They aimed positively charged alpha particles at gold foil. The experiment showed that:

- most alpha particles passed straight through the gold foil

- some alpha particles passed through the gold foil, but changed direction

- a very small number of alpha particles were reflected by the gold foil.

The diagram below shows four alpha particles, A, B, C and D, approaching four gold atoms in the foil.

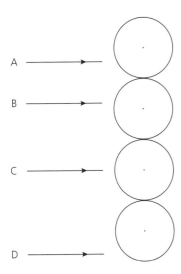

a) Complete the diagram to show the paths that the four alpha particles follow as they pass through the gold atoms. [4 marks]

b) Explain how the measurements led to the conclusion that the mass and positive charge of an atom are concentrated into a small nucleus. [4 marks]

....

....

....

....

....

....

(Total = 8 marks)

14 The pie chart shows the main sources of background radiation in the UK.

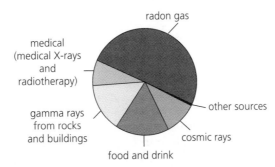

For any **three** types of background radiation shown on the chart, name the type and suggest **one** way to reduce the exposure to that source of background radiation. [6 marks]

Type 1: ....

How to reduce exposure:

....

....

Type 2: ....

How to reduce exposure:

....

....

Type 3: ....

How to reduce exposure:

....

....

(Total = 6 marks)

15 The half-life of uranium-238 is 4.47 billion years. Uranium-238 eventually decays into lead-206. Older rocks contain more lead and less uranium than younger rocks. The amounts can be measured and used to find a rock's age.

Suggest the advantages and disadvantages of using half-life to calculate the age of rocks. *[5 marks]*

....

....

....

....

....

....

....

(Total = 5 marks)

16 A nuclear power station is used to generate electricity. The power station includes fuel rods, control rods, moderator, boiler and water circulating in pipes. Which part of the nuclear power station is used to control the rate at which electricity is generated? Circle the correct answer. *[1 mark]*

A control rod

B moderator

C fuel rod

D boiler

(Total = 1 mark)

8 Astrophysics

1   Which of these correctly lists the objects in order of increasing size?
    Circle the correct answer.   *[1 mark]*

    A   moon < star < planet < galaxy

    B   comet < galaxy < star < universe

    C   moon < planet < galaxy < universe

    D   planet < star < comet < galaxy

    (Total = 1 mark)

2   Gravity causes some objects to orbit other objects. Which of these objects is not orbiting
    another object? Circle the correct answer.   *[1 mark]*

    A   moon

    B   star

    C   asteroid

    D   comet

    (Total = 1 mark)

3   The Moon takes about 28 days to orbit Earth once. The radius of its orbit is 385×10^6 m.
    Calculate the Moon's orbital speed.   *[4 marks]*

    ....

    ....

    ....

    ....

    ....

    ....

    (Total = 4 marks)

4   Halley's comet is a famous comet that orbits the Sun with a period of approximately
    76 years. Describe **one** way in which a comet is similar to Earth and **one** way in which
    it is different.   *[2 marks]*

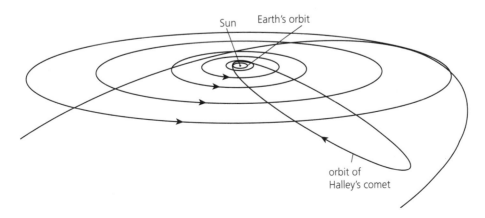

Sun   Earth's orbit

orbit of
Halley's comet

....

....

....

....

(Total = 2 marks)

5 a) Describe the evolution of a star much more massive than our Sun, from the nebula
   stage in its life cycle.   *[4 marks]*

....

....

....

....

....

....

b) Stars are stable during the main sequence stage of their life. Explain how forces in a
   star such as the Sun keep it stable during this stage in its life cycle.   *[3 marks]*

....

....

....

....

c) Stars can be classified according to their colour. Explain, in detail, why the colours
   of a red dwarf star, a main sequence star and a white dwarf star are different. *[4 marks]*

....

....

....

....

....

(Total = 11 marks)

6 A Hertzsprung–Russell diagram places stars in different positions according to their luminosity and surface temperature. Stars in different stages of their life cycle have different luminosity and surface temperature. Which of the table rows correctly shows how luminosity and surface temperature vary according to the life cycle stages of a star? Circle the correct answer, choosing from A to D.   *[1 mark]*

	Low temperature, high luminosity	Low temperature, low luminosity	High temperature, high luminosity	High temperature, low luminosity
A	main sequence	main sequence	red supergiant	white dwarf
B	white dwarf	red super giant	main sequence	main sequence
C	main sequence	main sequence	white dwarf	red supergiant
D	red super giant	main sequence	main sequence	white dwarf

(Total = 1 mark)

7 An astronomer is observing two stars. One star has an absolute magnitude of 1, and a more distant star has an apparent magnitude of 1.

State the difference between absolute magnitude and relative magnitude, explaining which star is brighter.   *[3 marks]*

....

....

....

....

....

....

(Total = 3 marks)

8 It is generally believed that the Universe began about 14 billion years ago from a single point in space and has been expanding ever since (the Big Bang theory).

Describe **two** pieces of evidence that support the Big Bang theory, explaining how they back up this idea.   *[6 marks]*

....

....

....

....

....

....

....

(Total = 6 marks)

9 a) Explain what is meant by the **Doppler effect**, giving **one** example. *[3 marks]*

....

....

....

....

....

Astronomers use spectroscopes to examine light from stars in the Milky Way galaxy, and from stars in the Whirlpool galaxy which is about 25 million light years away. Line spectra for light from the Whirlpool galaxy have been red shifted.

near galaxy

distant galaxy

violet ⟶ red

increasing wavelength

b) Explain what the term **red shift** means. *[2 marks]*

....

....

....

c) Describe how a line spectrum that has been red shifted looks different from a line spectrum that has not been red shifted. *[3 marks]*

....

....

....

....

d) Explain how the red shift provides evidence for the expansion of the Universe and the Big Bang theory. *[4 marks]*

....

....

....

....

....

....

(Total = 12 marks)

10 a) Write down the equation connecting wavelength, change in wavelength, velocity of a galaxy and the speed of light. *[1 mark]*

....

....

b) Light of wavelength 505×10^{-9} m is emitted from stars in a galaxy. The galaxy is moving away from Earth at 4000 km/s. Calculate the change in wavelength for this light. *[3 marks]*

....

....

....

....

(Total = 4 marks)

Notes

INTERNATIONAL
GCSE
(9–1)

Physics

for Edexcel International GCSE

Maximise your performance with practice questions, written to support and enhance the content of the Edexcel International GCSE Physics book.

◆ Enhance learning with extra practice designed to support the Student Book.
◆ Test knowledge with a variety of exam-style questions including multiple choice.
◆ Perfect for homework and independent study to ensure you have understood concepts.

HODDER EDUCATION

www.hoddereducation.co.uk

ISBN 978-1-5104-0566-0